EXPLORING ENGINEERING CAREERS

Exploring MECHANICAL ENGINEERING Careers

Andrea C. Nakaya

ReferencePoint Press®

San Diego, CA

ReferencePoint Press®

Printed in the United States

For more information, contact:
ReferencePoint Press, Inc.
PO Box 27779
San Diego, CA 92198
www.ReferencePointPress.com

LIBRARY OF CONGRESS CATALOGING-IN-PUBLICATION DATA

Name: Nakaya, Andrea C., 1976- author
Title: Exploring Mechanical Engineering Careers / by Andrea C. Nakaya.
Description: San Diego, CA : ReferencePoint Press, Inc, 2027. | Series: Exploring Engineering Careers | Includes bibliographical references and index.
Identifiers: LCCN 2025048176 (print) | LCCN 2025048177 (ebook) | ISBN 9781678212889 library binding | ISBN 9781678212896 ebook other
Subjects: LCSH: Mechanical engineering--Vocational guidance--Juvenile literature
Classification: LCC TJ157 .N35 2027 (print) | LCC TJ157 (ebook) | DDC 621.023--dc23/eng/20251222
LC record available at https://lccn.loc.gov/2025048176
LC ebook record available at https://lccn.loc.gov/2025048177

CONTENTS

Mechanical Engineer at a Glance

Working Conditions

Mostly in an office or laboratory; occasional visits to worksites such as factories and construction sites

Licensing

License required for engineers overseeing other engineers or providing services directly to the public

Personal Qualities

- Creative
- Problem solver
- Works well with other people
- Adaptable
- Analytical
- Curious

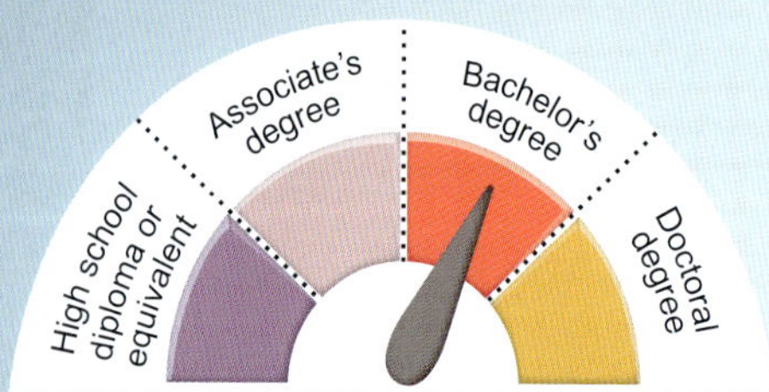

Minimum Educational Requirements

Bachelor of science in mechanical engineering or a related field

Pay $102,000 median salary

Future Job Outlook

Growth through 2033

11%

291,900

Number of jobs

Bureau of Labor Statistics, "Mechanical Engineers," *Occupational Outlook Handbook*, August 28, 2025. www.bls.gov.

Creating a Better Future

On the career information website CareerVillage, engineer Stephen Griffin writes about the ten years he has spent in engineering things. When it comes to the field, he says, "One of the first thing I think of . . . is the variety in the type of work you can do, the myriad of roles you can undertake and the wide array of industries you can work in. My own career has seen me delve into product design, quality, manufacturing, technical sales, and continuous improvement. I've worked in the fields of aerospace, automotive, and consumer goods."[1]

As Griffin's comment illustrates, the field of engineering includes a variety of jobs. Different types of engineers design bridges and buildings; develop cell phones and household appliances; work with chemicals; and build ships, cars, and airplanes. Others make textiles, improve manufacturing processes, develop electrical and lighting systems, and create robotics systems. While their roles within these industries differ greatly, all engineers are problem solvers who use math, science, and creativity to design a wide variety of technologies, structures, systems, and devices that people use in their daily lives.

Essential and Highly Valued Members of Society

Engineers are responsible for so many of the products that people depend on that it is hard to imagine life without them. The University of British Columbia Engineering Department explains on its website that evidence of engineers' work is all around: "Every aspect of our built environment has been touched by an engineer. From our roads and bridges to manufacturing processes, from the way vaccines are delivered and biomedical advances like prosthetics and laser eye surgery, and from pollution mitigation measures to our ability to recharge batteries . . . all these things came about because of an engineer."[2]

These professionals are highly respected in most communities because the work they do makes a difference to society and helps create a better world. According to the American Society of Mechanical Engineers' code of ethics, one of the fundamental obligations of engineers is "using their knowledge and skill for the enhancement of human welfare."[3]

> "Every aspect of our built environment has been touched by an engineer."[2]
>
> —University of British Columbia Engineering Department

Public surveys consistently show a high level of public respect for engineers. For instance, Gallup has been conducting opinion polls since 1976 on various professions, and the responses reveal that most people believe engineers to be honest and ethical. In its 2023 poll, engineers were ranked the third most trusted of twenty-three professions. Again, much of that opinion is informed by a widespread belief that engineers have the best interests of society at heart, working to advance the health, safety, and well-being of people everywhere.

Mechanical Engineering

The engineering field is typically divided into five main branches: chemical, electrical, civil, industrial, and mechanical. Each of these branches includes many additional specialized subfields. In general, though, chemical engineering is focused on chemicals and chemical processes, electrical engineering on electrical equipment and systems, civil engineering on infrastructure such as roads and bridges, and industrial engineering on industrial processes.

The last branch, mechanical engineering, involves designing, creating, testing, and maintaining systems and devices. It is one of the broadest branches of engineering, with a long list of applications. Mechanical engineer Wassee Syed comments that the practice and principles of mechanical engineering have been around for as long as people have existed. He says, "In some ways, 'mechanical engineering' has been around longer than any other form of engineering. The stone age man designed tools for hunting, and devised techniques for survival. Most of this activity could be classified as Mechanical Engineering."[4] In the present, mechanical engineers work with airplanes, cars, energy systems, biomedical devices, manufacturing processes, robotics, nanotechnology, and more.

Changing the Future

Through the things that they create, mechanical engineers not only make a difference in the way people live in the present, they also help shape the future. Yassmin Abdel-Magied, a mechanical engineer for Shell, maintains that society will continue to need mechanical engineers for their ability to help make life better. She says that in the future, "it will be mechanical engineers that debate and solve the most significant problems the world is facing. From tackling climate change, to overcoming water and food shortages, I believe we have the mix of creative, analytical and practical skills needed to meet these huge challenges."[5]

CHAPTER ONE

What Does a Mechanical Engineer Do?

Mechanical engineering is one of the broadest fields in engineering. Because mechanical engineers work in almost every industry and perform so many different types of tasks, it might be difficult at first to define what it means to be a mechanical engineer. The Institution of Mechanical Engineers helpfully explains that all mechanical engineering roles involve working with moving parts. It states, "Mechanical engineering is all about movement. It is the application of science and technology to solve real-world problems, through the manipulation of moving parts."[6] Mechanical engineers design, create, and repair devices and systems that have moving parts. They use engineering principles such as force and motion to guide their work. For this reason, they are instrumental in the design and operation of cell phones, cars, heating and air-conditioning systems, household appliances, machinery, medical devices, robotics, and many other products that meet people's daily needs and improve their lives.

Analysis and Design

Analysis and design are central to any mechanical engineering project and are a major part of what mechanical engineers do. Most engineers start a project by conducting analysis. This means studying a particular problem or an existing device or system to understand how it works and how a new or better product can be created. For example, mechanical engineers who are trying to design a car engine that uses less fuel might research existing engines to understand how current technology works and where they might be able to make improvements. They may also assess things like size and weight limitations, materials costs, and safety regulations in order to understand any constraints on their work.

> **"Mechanical engineering is all about movement. It is the application of science and technology to solve real-world problems, through the manipulation of moving parts."[6]**
>
> —Institution of Mechanical Engineers, a global engineering organization

After assessing what is needed, the successes and limitations of existing products, and any constraints, the mechanical engineers create a design. They might start by brainstorming many potential solutions, but eventually they will narrow it down to the best design and will create precise technical drawings to detail that design. Most engineers use computer-aided design software for this step. Using those technical drawings, a prototype of the design can be created and tested. By testing, a design can be evaluated to see whether it works as planned and what type of improvements are needed.

The initial design is rarely the end of the engineering process. Instead, most mechanical engineering projects involve repeated testing, evaluation, and redesign in what is often described as an iterative process. In a blog post, engineering company SGW Designworks explains that continued analysis and design are key elements of engineering. It states, "Engineers repeatedly evaluate and refine their designs, learning from each test and adjusting their approach accordingly. This approach ensures that the final product not only meets the initial design objectives but also addresses any unforeseen issues or challenges that may have emerged during the development." The company adds that this process is not something that can be accomplished alone. Instead, it says, "the process is also highly collaborative, often requiring input from a diverse set of disciplines. Engineers work alongside stakeholders, customers, and other professionals to gather requirements, generate design criteria, and ensure that the solutions are viable within the given constraints."[7]

Mechanical Engineering Job Fields

Because the field of mechanical engineering is so broad, most mechanical engineers specialize in a specific type of design. There are hundreds of different ways to specialize across dozens of different industries and in specific positions within those industries. Mechanical engineer and cofounder of the aerospace company Right Ascension 231 Zozscha Bomhardt says, "There are so many career options open to Mechanical Engineers. It feels like no two career paths are the same. . . . There's literally something for everyone. And you can choose a path that is more technical and less managerial or vice versa, whatever fits your style, personality, preferences, and goals."[8]

Mechanical engineers design, create, and repair devices and systems that have moving parts. They rely on engineering principles such as force and motion to guide their work.

> "There are so many career options open to Mechanical Engineers. It feels like no two career paths are the same. . . . There's literally something for everyone."[8]
>
> —Zozscha Bomhardt, vice president of operations at the aerospace company Right Ascension 231

Automotive Engineering

Mechanical engineers who work in automotive engineering try to improve the design and performance of vehicles. They work on many types of moving vehicles, from cars to buses, and from traditional to alternative fuel systems. There are numerous specializations within this field, including aerodynamics, suspension, safety, fuel systems, high-performance vehicles, wheels, reliability, and automation. Hironori Chaya works in automotive engineering at Toyota, as part of a team that is focused on safety and uses simulators to test the effectiveness of braking systems. He says, "My team is developing technologies to realize a society where there are no traffic accidents and where everyone can move safely and confidently."[9]

Thermal Engineering

Thermal engineering is focused on creating and managing heating and cooling systems. This includes air-conditioning, ventilation, refrigeration, and heating. Engineers in this field might work on systems for homes, schools, retail spaces, vehicles, or airplanes, focusing on maintaining comfort, safety, and efficiency. Thermal engineering is also used in product design. For instance, thermal engineer Malcolm Gutenburg works at NVIDIA, a company that makes computer chips and systems. These products generate heat as they operate, and Gutenburg's job is to figure out ways to dissipate that heat so that the products do not get too hot. At the same time, he needs to make sure the products have as much computing power as possible. He explains, "Our products constantly require innovation to keep increasing their capabilities and efficiency."[10]

Aerospace and Defense Engineering

Mechanical engineers who work in aerospace and defense develop aircraft, spacecraft, and defense technology such as drones, missiles, and satellites. Similar to automotive engineers,

Mechanical Engineering for NASA

Kate Gunderson is a mechanical engineer who has been employed at the National Aeronautics and Space Administration (NASA). One of her first jobs there was working on the Johnson Space Center's Gulfstream aircraft, which are used to fly astronauts back to Houston after they have returned from the International Space Station. She explains that these aircraft were originally designed to transport only two astronauts but needed to accommodate up to four crew members plus any other essential personnel, such as a flight doctor, to check the crew's health. She reconfigured the passenger cabin entirely. She explains:

> I used existing passenger seating to create the base for mattresses that are installed so they have a place to lay down, mounted medical oxygen bottles under each bed, ensured access to medical-grade outlets for special equipment, selected the color of new carpeting to be installed, and installed curtains for privacy around each bed. Yes, I somewhat jokingly, yet also seriously now consider myself an amateur aircraft interior designer. If you can believe it, I found space for four beds and six additional passengers plus two pilots, a Flight Science Officer and [one more person]. Whew, that was tricky!

Kate Gunderson, "My First Project as a NASA Engineer: DReAM," *The Plane Kate* (blog), May 2, 2021. https://katelyngunderson.com.

they focus on many different elements, including design, safety, and efficiency. They may also specialize in one part of an aircraft or other product. Jocelyn Mendoza is a mechanical engineer who works for Northrop Grumman, an aerospace manufacturing company. She specializes in the fuselage of the F-35 jet and explains that she enjoys seeing how smaller parts like this can turn into something big like an entire jet. She says, "It makes me proud to work on such a project."[11]

Biomedical Engineering

Biomedical engineers apply engineering concepts to the human body. They create solutions for medical issues and design devices that help improve people's quality of life. For example, they create artificial heart valves, pacemakers, artificial limbs, prosthetics, and implants. They also create medical tools like diagnostic systems and surgical robots. Biomedical engineer Kalifa Clarke works at a medical device company called ConMed, where she helps develop implants and surgical instruments that are used to help repair knee injuries. She says, "I love that . . . the products I work on help people." She explains that one interesting part of her work is evaluating prototypes. "Once I finish working on a prototype, I have surgeons test it out on a cadaver. . . . In the cadaver lab, the surgeon can test the product on a cadaver to see whether the new product or technique works. I receive valuable feedback on the function of the prototype as well as lessons on the anatomy and procedures."[12]

Nanotechnology

Nanotechnology involves manipulating atoms and molecules, which are invisible to the naked eye. It can be used to create new materials and devices and has applications in many different fields, including energy, health care, automotive, and aerospace. Engineer Gabriel Cossio is the creator of Nanoscale Labs, a company that aims to use nanotechnology in ways that help protect the environment. He talks about producing environmentally friendly fishing lures with this technology, explaining, "It turns out that fishing lures are extremely toxic and a main source of aquatic pollution in American fisheries."[13] Some of that danger comes from the use of volatile organic compounds to make traditional lures colorful. In contrast, he says that his company has implemented nanopatterning to replicate vibrant coloring without using these compounds. In addition, he says nanotechnology allows him to create lures from nontoxic polymers that are stronger and even have self-healing properties.

Anticipating Future Needs

Nanotechnology, biomedical technology, aerospace and defense, thermal systems, and automotive are just some of the most common branches of modern mechanical engineering. Mechanical engineers also work in manufacturing, construction, materials science, energy, acoustics, ergonomics, and many other fields, and there is just as much variety in their roles within those fields today.

However, mechanical engineers do not just solve existing problems. Instead, they often think ahead to the future needs of society, trying to anticipate the problems people will face in the future and working to develop solutions before those problems become urgent. By doing so, they can reduce any negative impacts on society. Examples of future-related challenges that mechanical engineers might think about are how cities can adapt to rising sea levels caused by climate change or how the growing global population will be able to feed itself. Looking forward helps mechanical engineering grow as a field, ensuring that this profession will always remain important to building a better world.

How Do You Become a Mechanical Engineer?

Writing on the University of Michigan Mechanical Engineering Department website, current and former mechanical engineering students talk about how they came to work in this field. Their stories reveal that there are many different paths to mechanical engineering. Some students say that they always knew they wanted to become a mechanical engineer. For instance, Asher Feigenbaum states, "For as long as I can remember, I have always loved understanding how things work, taking stuff apart, and putting them back together. From the age of 3, I was building structures of blocks taller than me. . . . This passion for creating was complemented by my favorite subjects of science and math, propelling me toward mechanical engineering."[14] However, other students say that it took longer to find mechanical engineering, and they report that even as they reached the end of high school, they had no idea what they wanted to do. Michelle Ruffino recalls, "I didn't always know that I wanted to be an engineer. Throughout middle and high school, my peers knew what they wanted to major in at college, and what career they wanted to have when they graduated. I, on the other hand, was unsure."[15]

The Importance of Education

As these stories show, the path to mechanical engineering is different for everyone. Regardless of how and when a person decides to become a mechanical engineer, though, anyone who wants to work in this field needs an engineering education. Mechanical engineers have a strong foundation in math, physics, and mechanics, and most earn at least a bachelor's degree in their field.

High school students who know they want to study mechanical engineering can begin building the knowledge and skills they will need by taking math, physics, chemistry, and computer science classes. A foundation in these subjects can be an advantage because it will likely make it easier to expand on these concepts in college. However, many people study engineering in college without having taken any related classes in high school. They simply have to learn the basics as part of their college curriculum.

Researching Schools and Majors

Research is another important part of the path to becoming a mechanical engineer. Mechanical engineer Wassee Syed advises that students interested in this career do as much research as they can, as early as possible. This includes talking to people who have experience in mechanical engineering. He says, "I think it is extremely important to gain as much information about your selected area of study as you can before you apply to it. Yes, you can look things up online. And yes, you can always change

Do Not Be Afraid to Try

When asked "What advice would you give to students thinking about studying mechanical engineering?," mechanical engineer Shelbi Tippett stresses that hands-on mechanical work is extremely beneficial. She advises that students not be afraid to try building or fixing things, even if they fail at first. Tippett says:

> No matter what you do in life, DON'T BE AFRAID. Gain some hands-on experience. Fix the leaky bathroom sink, change the oil in your car, build some Ikea furniture. It may take the internet and a couple of tries but DON'T BE DISCOURAGED! Trying is the only way that you become comfortable investigating how things work and eventually figuring out how to fix or redesign them. Once you establish this skill set and associated CONFIDENCE, the rest of it—completing the coursework, finding an internship, choosing a career path, etc.—it will all just fall into place.

Quoted in University of Maryland, Baltimore County, Department of Mechanical Engineering, "ME Careers." https://me.umbc.edu.

majors after you have applied. But in my opinion, nothing beats direct advice from those who are qualified to give it. It might save you a lot of time and frustration."[16] Students can also research colleges that offer mechanical engineering programs and even visit the schools that they are interested in, if possible.

Earning a College Degree

Most mechanical engineers have a bachelor's degree in mechanical engineering or a related field. Students should look for an engineering program recognized by the Accreditation Board for Engineering and Technology, since according to the Bureau of Labor Statistics (BLS), most employers look for employees from an accredited program. Possible mechanical engineering degrees include a bachelor of science in mechanical engineering, a bachelor of engineering, a bachelor of science engineering, and a bachelor of technology. Depending on the specific school, students may be able to choose a specialization within that degree, such as automotive engineering, renewable technology, or robotics.

> **"Mechanical engineering is incredibly broad as we take classes in statics and dynamics, thermodynamics, heat transfer, fluid mechanics, strength of materials, design, control systems, materials, and more."[17]**
>
> —Peyton Erickson, mechanical engineering student at the University of Southern California

Mechanical engineering students generally start by learning foundational math, physics, and chemistry classes. After that, they study a wide range of engineering principles. Peyton Erickson, a senior studying mechanical engineering at the University of Southern California, describes some of the many courses involved in a mechanical engineering degree. She says, "Mechanical engineering is incredibly broad as we take classes in statics and dynamics, thermodynamics, heat transfer, fluid mechanics, strength of materials, design, control systems, materials, and more. The breadth of the major is one of my favorite things about mechanical engineering."[17]

After getting their bachelor's degree, some students go on to earn an advanced degree, such as a master's or even a doctorate. This can lead to more job opportunities and higher-paying positions. It can also prepare students to teach, if that is their goal. Some schools offer combined programs that allow students to earn both a bachelor's and a master's degree in approximately five years. Advanced degrees typically require students to participate in research, such as a project or thesis.

Studying Mechanical Engineering Can Be Challenging

Nearly every mechanical engineering student agrees that studying the subject in college is hard work. Mechanical engineer Lauren T. Buckler says, "The time I spend in undergraduate wasn't easy; the work load was frustrating compared to my non-engineering roommates and friends."[18]

However, most students also say that while completing their studies took dedication, their degree was worth it. For instance, Buckler says, "I had a job before I graduated and it was a job I liked. A degree in Engineering opens a variety of career options; there are opportunities to find a career you enjoy."[19] University of Michigan mechanical engineering student Beth Ann Less insists that the hard work made her better at her job. She says, "While I definitely didn't enjoy the long nights camped out in the library or the intentionally vague instructions that accompany most lab projects. . . . Looking back, I now recognize that overcoming these difficult challenges [in school] has made me a stronger and more capable problem solver and engineer."[20]

Gaining Practical Experience

In addition to getting an education, students who want to become mechanical engineers need to accumulate practical experience. This will allow them to apply their engineering skills to real-world projects. Many people advise trying to get experience while still in school. For younger students, that can simply mean working on a project related to mechanical engineering. Jon Van Noord is chief mechanical engineer in research at the Space Physics Research Laboratory and XTRM Labs. He says, "If you want to

Preparation for a mechanical engineering career includes both education and practical experience. Participation in a high school robotics program teaches real-world engineering skills.

be an engineer try and get practical experience building something as early as possible. It might be on a FIRST Robotics team in grade school or high school, or it could be their own personal projects or on a student team."[21]

Older students can look for more formal volunteer or job opportunities related to mechanical engineering. Mechanical manufacturing engineer Brian Calbeck explains that early jobs might not seem exciting, but they can eventually lead to something better. He says, "Look for unique jobs and experiences that interest you and relate to your course of study. Your first internship probably won't be your dream one. I spent two summers working in a machine shop. I started out sweeping the floors but worked my way up to running the machines." Calbeck and others explain that the more experience students have, the better their chances of finding a job after they finish school. Of his college internship, Calbeck says, "That job not only gave me great

Do Engineering Students Have a Social Life?

The EngineerGirl website is designed to educate girls and women about opportunities in engineering. In one post, it addresses a common fear that engineering students are too busy to have a social life in college. Mechanical engineering graduate Claire Elliott responds:

> I did have a social life in addition to the heavy workload, but my social and work life were connected. In my sophomore year, I found ASME (American Society of Mechanical Engineers) and joined. This club gave me a way to make friends that were taking the same courses I was. We formed study groups, and the older students were free with advice of which professors to take and avoid. These were the friends that I hung out with on weekends, went to parties with, studied with, shared apartments with, etc. The club itself hosted lots of activities, trips to conferences around the country, weekly meetings where I made connections with engineers from the industry, and mentorships with professors. . . . The bottom line is that if you're interested in engineering, and you're willing and capable of doing the work, go for it. Be prepared to put in the work, but the social life will be there, too.

Quoted in EngineerGirl, "Do Engineering Majors Have a Social Life?," July 26, 2021. www.engineergirl.org.

hands-on experience but helped me stand out when I went to apply for opportunities later on in my college career."[22]

Experience is also valuable because it can help students figure out what area of mechanical engineering they want to focus on. Many mechanical engineers work in multiple fields before they find the best fit. Warren Seering, a professor of mechanical engineering at the Massachusetts Institute of Technology, says that it is common for young mechanical engineering graduates to jump from job to job before finally settling into one position. For this reason, he stresses that new graduates should look for a job where they can try new things. "Find a company that lets you continue to learn," he says, "On the first job, that's more important than location or salary."[23]

Licensing Requirements

Many entry-level jobs do not require licensing, but to work more independently or in a leadership role, a mechanical engineer needs to obtain a license. According to the BLS, any mechanical engineer who wants to oversee the work of other engineers or to provide services directly to the public must be licensed. Licensing is offered through the National Society of Professional Engineers and is intended to help ensure public safety. A licensed engineer is commonly referred to as a Professional Engineer (PE).

To earn a license, an engineer must complete a four-year college degree and also work for at least four years under the guidance of another PE. There are also two licensing exams that must be taken: the Fundamentals of Engineering exam and the Principles and Practice of Engineering exam. The first exam can be taken after graduating from college, while the second should be taken after fulfilling the work experience requirements. Finally, while state requirements vary, some states require ongoing education in order to renew this license.

Ongoing Education

Earning a degree and obtaining a license is not the end of the learning process for mechanical engineers. Individuals who work in this field continue learning throughout their professional careers. The American Society of Mechanical Engineers (ASME) explains that continual learning is necessary because the field of mechanical engineering changes quickly. The organization explains, "As mechanical engineers, our field is constantly evolving with new technologies, techniques, and industry trends. To stay ahead in this dynamic landscape, embracing lifelong learning is not just advantageous but essential."[24] In its code of ethics, the ASME explicitly states, "Engineers shall continue their professional development throughout their careers."[25] Such ongoing education can take many forms. Some engineers spend time learning on their own, while others attend

"As mechanical engineers, our field is constantly evolving with new technologies, techniques, and industry trends. To stay ahead in this dynamic landscape, embracing lifelong learning is not just advantageous but essential."[24]

—American Society of Mechanical Engineers, professional organization for mechanical engineering in the United States formal classes. In addition, learning can happen at workshops, conferences, and other industry events.

Continuing professional development helps mechanical engineers stay informed about best practices and innovations in the field. By staying current with emerging technologies and industry standards, mechanical engineers can design safer, more efficient, and more sustainable solutions that advance the well-being of their communities and society.

CHAPTER THREE

What Skills and Personality Qualities Matter Most—and Why?

When choosing a job or career field, it is important for potential employees to determine whether their skills and personality are a good match for that role. The Chartered Institute of Personnel and Development, an association for human resources management professionals, explains why: "When a person's skills are well matched to their role, it shines through in greater job satisfaction and productivity. But when there is a skills mismatch . . . it's bad news for the individual and the employer. Skills mismatches often manifest in trust issues, lower job satisfaction, lower productivity, lower confidence and higher stress."[26] To make sure that they are a good match, aspiring mechanical engineers should understand what skills and personality qualities are most important in this field.

Math Is Essential

Math is one of the most vital tools in mechanical engineering. Engineers utilize calculus, statistics, differential equations, and other types of math on a daily basis to analyze problems and create effective designs. For example, math helps them understand how things move, how objects will act under various forces, and how movement will change over time. It is impossible to be a mechanical engineer without math skills.

While math is important, not all mechanical engineers naturally excel at it. Some say that math was not their best subject in school, and others explain that they started out without strong math skills but were able to learn what they needed over time. Frank Harwath, director of the engineering program at North Central College in Illinois, insists that even though math is important, if a student has struggled with math in the past, they should not rule out a career in engineering. With practice, almost anyone can learn the math skills needed to become an engineer, he says: "We all can do math and science. Really! . . . The more you use these tools the more familiar they become."[27]

The University of the Pacific website explains that engineering students who need extra help learning math are likely to have many resources available to them at school. The university states, "If you think math is difficult, you should seek extra help. You'll want to speak with your teachers, who may be able to give you extra support. Your school may also have staff or peer tutors who can help you crack those tougher concepts. . . . Most universities also have academic support centers that offer drop-in or scheduled tutoring, taught by high-achieving peers or graduate students."[28]

Physics and Mechanical Skills

Physics is another foundational skill for mechanical engineering. Physics is a fundamental set of principles that describes how matter behaves and moves. Mechanical engineers use these principles to understand how objects move and how they respond to different types of forces. For instance, a mechanical engineer designing an aircraft can use physics to understand the different types of forces—like lift and thrust—that keep an aircraft aloft. Like math skills, an understanding of physics is essential for a mechanical engineer. Discover Engineering, a website that provides information about engineering, insists, "Physics . . . gives mechanical engineers the basic rules and ideas to work with."[29]

> "Physics . . . gives mechanical engineers the basic rules and ideas to work with."[29]
>
> —Discover Engineering, a website that provides information about engineering

Mechanical engineers also need mechanical skills, meaning the ability to physically work with mechanical systems. They need to understand how these systems and their associated tools work and how to manipulate them. For example, a mechanical engineer who specializes in kitchen appliances needs to understand how these appliances work and how they are assembled and disassembled in order to be able to repair or improve them.

Importance of Problem-Solving

Problem-solving is central to any type of mechanical engineering job. It is fundamentally what mechanical engineers do; they break problems down and use their engineering knowledge and skills to come up with solutions. Engineer Jenny Boothby uses her problem-solving skills to design smart plastics at the Johns Hopkins Applied Physics Lab. She says, "I love problem solving! . . . My passion is coming up with new ideas for problems and being an engineer has always been the right choice for that."[30]

Mechanical engineers often approach problem-solving in a systematic way. They usually start by defining the problem so that they can fully understand it. For a complex problem, they might break it down into smaller pieces. After that, they gather information and brainstorm ideas for possible solutions. Finally, they can choose the best solution and test it by using it to solve the problem.

Creativity

Creativity is another important skill in mechanical engineering. Successful engineers need to be able to use their imagination to come up with innovative ideas and solutions. They also need to be flexible and open to new ideas. A Michigan Tech College of Engineering article explains, "There is never just one way to solve an engineering design challenge; there is no single, 'right' answer to a problem. . . . Engineering requires a sense of vision that goes

The Importance of Perseverance

Meagan has been working in mechanical engineering for seven years and says that this field is very rewarding. However, she also stresses that aspiring engineers should be ready to persevere in the face of challenges. She explains:

> Engineering is arduous, which if you are currently studying it you already know. Failure is also part of the job. Therefore, persistence is paramount. You could be the most talented engineer, but if you do not have determination, you probably will not be successful. The good news is that you are 100% in control of this one. Also, it does not require extra studying to get this skill, only heart!

Meagan, "Mechanical Engineering Skills: 23 Things You Need to Learn," Engineering Expectations, 2025. https://engineeringexpectations.com.

beyond constraints to 'see' a solution that others might miss or dismiss as farfetched."[31]

Creativity and open-mindedness not only help engineers solve daily work challenges, they can also lead to new career opportunities. "I'm still a work in progress," states engineer Pallavi Baddam. She shares that this optimistic and adaptable outlook has contributed to her success. "My career progression is the result of being open-minded at every turn, rather than pursuing a rigid path," she says. "From my first job onwards, I found that being flexible, and endlessly curious, created new opportunities and accidental career pivots I couldn't have foreseen."[32]

Be Able to Accept Criticism

Mechanical engineers also need to be open to accepting constructive criticism. Mechanical engineer Meagan explains that people who work in this field will constantly have their work reviewed by bosses, senior engineers, or peers. She says that this review is important because it ensures that important details do not get missed. However, she cautions that the constructive criticism that often comes from this review can sometimes make people feel bad about their work. Instead of feeling attacked, she advises looking on critique as a learning opportunity. She says,

"Know that the feedback is not personal and welcome it. The constructive criticism will make you a better engineer because you will know what to do the next time you have that task and be glad someone in your work caught the issue before the product launches."[33]

Interpersonal Skills

When it comes to the skills needed to be a mechanical engineer, many people think about technical skills like math and physics; however, mechanical engineers also need strong interpersonal skills, such as communication, teamwork, leadership, and conflict resolution. This is because, while engineers often spend a lot of time working alone, most also work as part of a team, and they need to be able to work well with that team. Mechanical engineering student Yee Siow stresses, "I . . . learned that being an engineer is so much more than being able to come up with elegant designs and smart solutions, and a big part of designing an outstanding product stems from good communication and teamwork."[34]

"I . . . learned that being an engineer is so much more than being able to come up with elegant designs and smart solutions, and a big part of designing an outstanding product stems from good communication and teamwork."[34]

—Yee Siow, mechanical engineering student

Communication skills are a vital part of almost any mechanical engineering job. A mechanical engineer may work as part of a team with other experts, or they may need to work with clients. In these roles they will need to gain an understanding of the wants and needs of others while effectively sharing their own ideas. In some cases the people they are communicating with may not be experts in mechanical engineering, making clear and effective communication even more important. Discover Engineering explains:

> You need to be able to explain complicated ideas in a way that's easy to understand and work well with people from different areas of expertise. . . . When you communicate

> well, you help keep projects on track and create a space where new ideas and solutions can grow. . . . Here's an example: Instead of saying "torque," you might explain it as "the force that makes something spin," which is much easier for most people to grasp. It's about making the hard stuff easy to get.[35]

Personality

A popular stereotype portrays mechanical engineers as introverted, meaning that they prefer to spend time alone. In contrast, extroverts thrive on social interaction. In reality, though, there are many of both personality types in this profession, and a person can be a successful mechanical engineer as either an introvert or an extrovert. Most mechanical engineering jobs require both focused individual work and effective cooperation and communication with others. Boeing engineer Jamie Krakover explains, "Engineering is a balance. There will be times where you work alone or in small groups, and have time to think through things before jumping in with [an] answer. There will also be times where you will be asked to give presentations of the work you've done and explain it to others." She stresses that nobody should have to change their personality to be a mechanical engineer because there is strength in that unique personality, insisting, "you can be a great engineer by being who you are."[36]

These are just some of the skills and personality qualities that matter the most in mechanical engineering. Other beneficial qualities include curiosity, adaptability, attention to detail, and analytical ability. While some people naturally have many of these helpful qualities and skills, students who lack some of them should not necessarily rule out mechanical engineering as a career. Through practice it is possible to develop many of these traits.

CHAPTER FOUR

What Is It like to Work as a Mechanical Engineer?

When asked what their typical day is like, many mechanical engineers are quick to reply, "It depends." Mechanical engineering jobs vary significantly depending on the engineer's experience, the field of work, the specific project, and the stage of that project. Yassmin Abdel-Magied is a mechanical engineer who is involved in drilling projects for Shell. She says, "The cool thing about a job in mechanical engineering is that there is no typical day."[37] Sometimes she collaborates with geologists to assess potential drill sites, and other times she is on an oil rig helping the drill supervisor get a job done.

> "The cool thing about a job in mechanical engineering is that there is no typical day."[37]
>
> —Yassmin Abdel-Magied, mechanical engineer for Shell

Where Mechanical Engineers Work

Mechanical engineers work in almost every industry, including automotive, aerospace, construction, health care, energy, manufacturing, defense, robotics, chemicals, and electronics. The Institution of Mechanical Engineers says, "It's one of the rare professions where you really can work anywhere."[38] However, some industries employ a lot more mechanical engineers than others. According to the BLS, there are just over 290,000 mechanical engineers in the United States. The agency finds that about 20 percent are employed in architecture, engineering, and related services. Approximately 15 percent are in machinery manufacturing, 10 percent in

Mechanical engineers work in any number of industries, including automotive, aerospace, construction, health care, energy, manufacturing, defense, robotics, electronics, and architecture.

transportation manufacturing, 7 percent in scientific research and development services, and 7 percent in computer and electronic product manufacturing.

The BLS reports that Michigan and California employ the most mechanical engineers. For metropolitan areas, Detroit has the highest employment level—at almost twenty thousand workers. Los Angeles and Boston are next, with about eight thousand jobs each. In most cases, large cities such as these have more mechanical engineering job opportunities than small towns or rural areas, because there are more industries there.

Work Environment

Most mechanical engineers spend the majority of their time working inside, in an office or laboratory. They typically spend a lot of hours using a computer to create, modify, and analyze designs.

Mechanical engineer Kyle Neher cautions, "Depending [on] what part of engineering you are in, you can get stuck at your desk a lot and get forced to stare [at] a computer for most of the day."[39] It is also common for mechanical engineers to travel to factories, construction sites, or other locations to inspect or fix mechanical equipment. For most, this constitutes a smaller percentage of their work time and depends on the industry that employs them. Some engineers spend a lot more time than others doing hands-on work at jobsites, and some often travel to where they are needed.

Whether they are in an office or traveling, most mechanical engineers work closely with other engineers or team members. Neher says, "I think the most surprising thing about being a mechanical engineer is how much we co-align and work with other departments in order to produce a final product. Producing a quality product relies on people with all kinds of backgrounds and educations to come together and work alongside each other to make the process run as smooth as possible."[40]

The pace of daily work and the stress associated with mechanical engineering varies. Many mechanical engineers say that they mostly work regular business hours and have a good work-life balance. However, some note that this type of work can involve stress and overtime. A lot of mechanical engineering work is project based, meaning that engineers work on specific projects, complete them, then move on to new projects. This can mean there are project deadlines that require periods of more intense work. Mechanical engineers may also work on multiple projects at once, which can be challenging. Other potential causes of stress in this profession are meeting budgets and the expectations of clients, as well as working on large, costly projects for which precision is crucial.

Mechanical Engineering Salaries

Mechanical engineering jobs typically pay well. As reported by the BLS, in the United States the median annual salary for a mechanical engineer is about $102,000. This is double the average

US salary, which is only about $50,000. When compared to the other major branches of engineering, it is about the same as the average salaries for civil and industrial engineering. Chemical and electrical engineers earn more—close to $120,000 yearly, according to BLS.

Within mechanical engineering, the BLS finds that scientific research and development is the top-paying industry, with a median annual income of about $123,000. Computer and electronic product manufacturing is the second-highest paying. Machinery manufacturing is the lowest-paying field within mechanical engineering, with an average salary of about $96,000. Earning potential also depends on location. For instance, according to the BLS, the District of Columbia, New Mexico, and California all have an annual median salary over $130,000. Some metropolitan areas—including Columbus, Georgia, and San Jose, California—pay above $150,000. However, the higher cost of living in these areas also influences the higher pay rate.

Daily Tasks

Malin is a senior mechanical engineer for Axis Communications. She works with cameras and sensors and says each day's focus can vary. "Depending on which phase of the project you are in, it can look quite different," she states. "In some periods there are a lot of meetings, in others we work with new concepts and then there is a lot of 3D modeling and evaluation of prototypes which takes up my time."[41] However, despite the variety, there are typical tasks common to most mechanical engineering jobs. Some of the things that mechanical engineers might spend their day doing include assessing what needs to be done and how to do it, coming up with ideas for new designs, using computer software to create detailed drawings of designs, working in a lab to create models or prototypes, and meeting with team members or clients to talk about design ideas, job requirements, or job progress.

Mechanical engineer Brandon Hayter says that he enjoys the variety of tasks involved and the fact that every day is different,

A Good Life

In the opinion of mechanical engineer Brandon Hayter, working as an engineer includes challenges, but overall, it is an enjoyable profession. He says:

> Most engineers would say that they have a good life. They get to solve problems and help make the world a better place. The engineering profession is highly respected, and many engineers I know feel they are making a difference in the world. Life as an engineer can be gratifying and satisfying. With a decent average salary, the life of an engineer is fun and full of adventure. I have been very fortunate to be able to travel, live in a beautiful city, and have enough time off work to enjoy friends and family. With the good, there are also many challenges that come with the job, but many of us feel that we are up for the challenge. We are constantly learning and growing as we work to make the world a better place. I am proud to be an engineer, and I believe that, in general, engineers have a good life.

Brandon Hayter, "My Daily Life as an Engineer," Peak Life, August 23, 2022. https://peaklifebrandon.com.

but he adds that this variety can also be challenging. He explains, "This [job] . . . requires both rapid task switching and uninterrupted deep work. The requirement for both of these is what I struggle with the most. At any one time, I may have 5+ customers trying to get a hold of me, with the same number of team members needing something, all while I am trying to concentrate and do deep work."[42] Hayter says that the key to dealing with this is to set aside uninterrupted work time and communicate to his colleagues that he is unavailable during those periods.

A Rewarding Career

While many mechanical engineers say that they often find their jobs challenging, most also say that this field is very rewarding. Mechanical engineering involves a lot of creativity. Some people describe it as a process of dreaming up ideas and then making them a reality, and this can be an exciting and fulfilling thing to do. Ben Truman, a mechanical engineer formerly with UTC Aerospace Systems, says that the desire to be a dreamer never goes away, insisting, "no matter what role you progress into, you can still be creative—from inventing new products to developing better processes."[43] Another thing that makes mechanical engineering rewarding is that many engineers feel like they are doing something that matters to people and makes their lives better.

A Male-Dominated Field

Aspiring mechanical engineers should be aware that men currently outnumber women in this field. In recent years, many schools have increased their efforts to encourage women to enter engineering, and data shows that the number of women is slowly increasing; however, the gap remains significant. According to 2024 BLS statistics, only about 11 percent of mechanical engineers are women.

> **"No matter what [mechanical engineering] role you progress into, you can still be creative—from inventing new products to developing better processes."[43]**
>
> —Ben Truman, mechanical engineer formerly with UTC Aerospace Systems

Despite gender differences, many women who do work as mechanical engineers stress that this is a fulfilling field. Mechanical engineer Erin maintains that while the ratio of women compared to men can be something to consider, it should not be the only thing that aspiring engineers focus on. She says, "For women pursuing engineering I'd say don't be intimidated by numbers differences between men and women, because when it comes down to it it's not about those numbers. If you have the skills or the drive and determination to pursue engineering then absolutely do that. It's an incredibly diverse field, in more ways than just between men and women, and everyone has something that they can contribute."[44]

Overall, most mechanical engineers say that they like their job. *U.S. News & World Report* ranks mechanical engineer at number twenty-seven in its list of "100 Best Jobs." It says that mechanical engineering jobs generally have a low level of stress, a high level of flexibility, and the opportunity for advancement. Together, these factors make mechanical engineering an appealing and rewarding career choice for many.

CHAPTER FIVE

Advancement and Other Job Opportunities

Career change is common in life, and most people work in multiple jobs over their lifetime. One BLS survey found that individuals held an average of twelve jobs between ages eighteen and fifty-eight. There are many reasons to change jobs. Some people want more challenge and responsibility; others desire higher pay or a different work environment. When it comes to changing jobs, mechanical engineers tend to have a lot of options because they possess a versatile and in-demand set of skills—including creativity, design, and problem-solving—that can be applied in a wide range of careers. Most are also experts in physics, thermodynamics, mechanical systems, and materials science, allowing them to move around in these fields.

Advancement Opportunities

After earning a bachelor's degree in mechanical engineering or a related field, a mechanical engineer will usually start out in an entry-level job. Entry-level jobs differ depending on the industry; however, this type of job often means working on a variety of tasks that help support senior engineers. That might include collecting data, running tests, creating drawings with computer-aided design software, and helping prepare documentation. Most engineers work in an entry-level position for at least a year, but with time, experience,

Transitioning into Design

Product design and development engineer Ron Llanes talks about how his mechanical engineering education made it easy to transition into the field of design. He says:

> Mechanical Engineering is such great career to get into because it is so versatile. . . . Going into college, I wanted to get into the fields of robotics or automotive. I did get a job after graduating in the automotive industry and I really enjoyed it. But after a down-sizing, I fell into the field of Product Design and Development. I never heard of this industry and I wasn't sure how my ME degree would translate. But the beautiful thing about a mechanical engineering degree is that you can use the principles and processes you learn in college in just about any field. I have now worked 30 years in this field, designing multiple products for companies big and small, and I owe it to the foundations I acquired as a mechanical engineer. As long as you keep open-minded and find what you are passionate about, that ME degree will serve you well."

Quoted in Career Village, "Mechanical Engineers, What Was the Reason You Choose Mechanical Engineering?," May 9, 2024. www.careervillage.org.

and dedication, many say that they quickly had the opportunity for advancement.

Mid-level mechanical engineering jobs include less supervision and more responsibility. Since they have more knowledge and experience, individuals at this level are able to work more independently. For instance, they might have their own projects, even though they may still need to reach out to senior engineers for help sometimes. Pay is also higher than for an entry-level position.

For some mechanical engineers, advancement eventually means moving into a supervisory role. In addition to engineering skills, these types of positions typically require communication, leadership, and teamwork skills. Bennett Wilson is a mechanical engineering manager at Markforged, a company that creates 3-D metal and carbon fiber printers. He says that his typical day is a mixture of technical work, managing projects, and supervising people. He explains that he likes to start the day by preparing for meetings and also focusing on technical work. After that, he says, there is a lot of teamwork and communication. He explains, "In broad strokes, the remainder of my day is spent ensuring we have clear direction and alignment between team members. This could mean designing experiments, holding design reviews, reviewing test results or clarifying priorities with the team."[45] Nathan is a mechanical engineering manager for an engineering company called MECAGINE. He says that his role includes training new team members. He explains, "We train newcomers on various subjects, in order to enable them to grow their skills. The topics covered can be technical . . . or the use of new software, manufacturing processes . . . [or] issues related to particular sectors of activity."[46]

Unique Job Opportunities

Another way to advance beyond an entry-level mechanical engineering job is to pursue a specialization in an area such as robotics, biomechanics, or nanotechnology. Engineers who specialize gain in-depth knowledge in their chosen field, which often makes them more competitive within the field and can lead to a higher salary. Specialization is common in mechanical engineering. At some schools, engineers can specialize as part of their undergraduate degree. They can also choose to pursue an advanced degree in their area of interest or earn a specialized certification.

Mechanical engineers who specialize may get to work in some unique jobs. Writing for the ASME, Cathy Cecere explains, "They work in situations that many in the general public will never have a chance to see or experience." For example, she says that some mechanical engineers get to test jet engines: "Engineers who evaluate jet [engines] at GE Aerospace's test site in Peebles, Ohio, get to throw chunks of dust, dirt, and ice at the monster machines."[47] Aerospace engineer Ben Truman talks about one of his unique job roles. He says, "I designed the first class cabin interiors for Emirates airlines in my first graduate role. Only a mechanical engineer could have found

> **"[Mechanical engineers] work in situations that many in the general public will never have a chance to see or experience."[47]**
>
> —Cathy Cecere, content program manager at the American Society of Mechanical Engineers

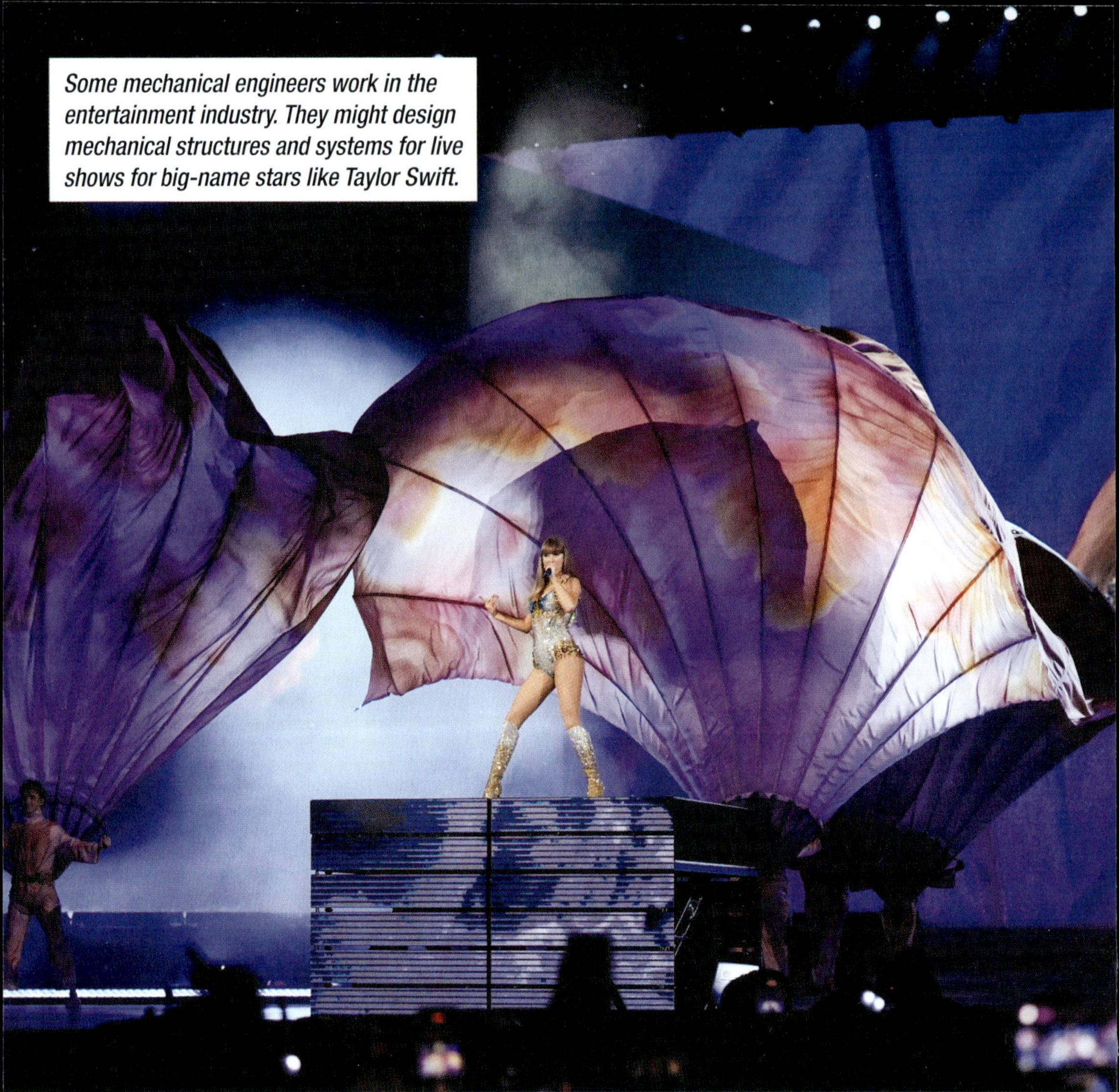

Some mechanical engineers work in the entertainment industry. They might design mechanical structures and systems for live shows for big-name stars like Taylor Swift.

a way to fit a fully-reclining chair, a desk, a wardrobe a minibar and a TV into such a tiny, yet stylish, space!"[48]

A writer for the University of California, Riverside, engineering blog talks about some other unique mechanical engineering roles. One is that of race car engineer, designing cars for Formula 1 racing. This job involves travel around the world. Sports engineer is another unique mechanical engineering role and involves improving products used by sports professionals, such as baseball bats. An entertainment engineer is yet another unique job role. This type of engineer designs mechanical structures and systems for entertainment venues. The blog author gives some examples:

"Giant revolving Cirque du Soleil stage? Check. Building the intricate sets for the next Hollywood blockbuster? You got it. Designing live shows for big-name rock stars like Taylor Swift and U2? All possible with a background in entertainment engineering."[49]

Career Transition

It is also common for mechanical engineers to say they were able to use their diverse set of skills to transition into a completely new type of position with relative ease; for instance, if the job market changes or they are not happy in their current role. Mechanical engineer Emilie insists that everyone should strive to find a job that is fulfilling and enjoyable. She says, "I am a real advocate for being brave and switching careers if you find you aren't in a field that is the right fit for you. . . . I really believe that you can find a career which is fulfilling and that you love waking up for everyday."[55] Fortunately, she says, engineering makes it easy to do this: "Engineering can open up so many doors, you can pivot into different sectors and job roles throughout your career."[51]

"Engineering can open up so many doors, you can pivot into different sectors and job roles throughout your career."[51]

—Emilie, mechanical engineer

Some mechanical engineers end up transitioning to a completely different field, like software development, consulting, or even entrepreneurship. Eight mechanical engineering professors recently collaborated on an article that was published in *Mechanical Engineering Magazine* to talk about what it means to be a mechanical engineer. They explain that the skills mechanical engineers gain during their education help set them up for many different careers:

> An ME [mechanical engineering] degree empowers graduates to pursue a remarkably rich range of career paths across a wide variety of industries as well as fundamental research. . . . While recent ME graduates often apply

their new knowledge and engineering practices to closely related fields, they can also leverage their critical thinking and problem-solving education in medicine, law, and other professions. For instance, at the universities where we teach, many graduates have gone on to lead businesses in industries as diverse as aerospace, automotive, and telecommunications; while others have become public communicators, high level lawyers, and astronauts. The skills learned by MEs are so important and widely applicable that two mechanical engineering graduates serve in the U.S. Senate.[52]

Networking

No matter what kind of job a mechanical engineer is interested in, networking can be an important part of reaching that goal. Networking is the process of building connections and relationships with other professionals. It could include talking to other engineers, finding a mentor, attending engineering events, or be-

Using the Principles of Mechanical Engineering in Firefighting

After earning a doctorate in mechanical engineering, Gavin Horn went on to work as a firefighter and then as research director for the Fire Safety Research Institute, where he aims to reduce the risks associated with fires. He says that his background in mechanical engineering has been invaluable in these jobs. He explains:

> Having a background in engineering built the foundation on how I approach a problem. In both research and emergency situations, this foundation helps me think clearly on my feet so I can bring the most positive outcome as the situation calls [for]. . . . My mechanical engineering education provided the background to understand fire dynamics, thermal dynamics, and heat transfer; it helps me to teach other firefighters to apply these to real life. . . . These skills are also useful to help with fire investigations to find the cause of each fire.

Quoted in Helen Chang, "Torching a Path from Engineering to Firefighting," Grainger College of Engineering, Mechanical Science & Engineering, University of Illinois, Urbana–Champaign, February 28, 2024. https://mechse.illinois.edu.

coming a member of a professional association like the ASME. Through networking, engineers can learn more about how the industry works and gain insight on how to be successful. They can also find out about job openings that might not be publicly advertised. Many engineers say that they utilize their professional network throughout their entire career.

Overall, mechanical engineers tend to have access to many job options. Whether they are looking for advancement or career change, it is common to hear them say that their versatile set of qualifications makes them marketable and allows them to take advantage of opportunities.

What Does the Future Hold for Mechanical Engineers?

It is difficult to imagine life without mechanical engineers. For so long they have been helping improve the way that people live, work, and play. Many of the products and systems that society relies on are designed by mechanical engineers, and when a new problem arises, it is often mechanical engineers who help find the solution. As a result, most experts see a strong future for this profession.

Future Demand for Mechanical Engineers

Over the next decade, demand for mechanical engineers is predicted to be high. The BLS predicts an 11 percent increase in mechanical engineering jobs through 2033, which is more than double the average US job growth rate of 4 percent. Overall, the BLS expects more than nineteen thousand job openings in this field each year over that period. It explains that one reason for growing demand will be an increase in automation, for which mechanical engineers will be needed to create and maintain computerized and robotic machinery and other products.

According to Cathy Cecere's article on the ASME website, in the United States, the supply of mechanical engineers is not sufficient to meet demand, making mechanical engineering a sought-after set of skills. Cecere explains that the main problem is that not enough people are becoming mechanical engineers. She says, "Graduates with majors in mechanical engineering are higher in volume than other engineering specialties—representing 25 percent of all engineering graduates. Yet, the volume of new talent coming through is less than a quarter of those sought by businesses." Further, she says that the current mechanical engineering workforce is aging, meaning there will be an even greater need for mechanical engineers in the future. She says, "With an average employee age of 43, just over the horizon is a cohort of engineers getting ready to retire."[53]

Sustainability and Renewable Energy

Experts predict that in the future, one of the highest areas of demand for mechanical engineers will be in renewable energy and sustainable technology. As Warren Seering, a professor of mechanical engineering at the Massachusetts Institute of Technology, bluntly explains, "We're running out of things. . . . We can't keep

Some mechanical engineers specialize in sustainability and renewable energy. In this field, they help design wind, solar, and hydroelectric power systems and develop ways to manage, store, and distribute the energy created by these systems.

using energy and materials the way we used to."[54] Mechanical engineers will be needed to come up with sustainable alternatives.

There are many ways to specialize within the field of sustainability and renewable energy. Mechanical engineers will help design and maintain wind, solar, and hydroelectric power systems that can provide energy without carbon emissions. They will also work on ways to manage, store, and distribute that energy. In addition, they will be needed to create energy-efficient and sustainable designs for homes and other types of buildings, and they will help develop electric vehicles and other types of transportation that do not rely heavily on fossil fuels. Other jobs in this field include developing carbon capture technologies that can reduce greenhouse gas emissions and providing solutions to climate change calamities, such as uncontrolled flooding, extreme heat, and rising sea levels. On the ASME website, independent writer Jerry Guerra insists:

"As the global community continues to address environmental challenges, the role of mechanical engineers will become pivotal."[55]

—Jerry Guerra, independent writer and consultant to engineering and architectural industries

> As the global community continues to address environmental challenges, the role of mechanical engineers will become pivotal. Their ability to innovate and implement sustainable solutions is essential for achieving energy efficiency, reducing emissions, and promoting environmental stewardship. . . . Mechanical engineers can lead the way in creating a more sustainable and resilient future.[55]

Artificial Intelligence

In recent years, artificial intelligence (AI) has become increasingly integral to many industries, and most experts believe that in the future, mechanical engineering roles will require a working knowledge of this technology. Mechanical engineers will adopt

AI to increase efficiency of systems, predict failure of devices, speed automation, and improve data analysis. While in some fields there are fears that AI will replace jobs, in mechanical engineering, most experts believe that that this tool will instead be used to help mechanical engineers work more efficiently and effectively.

Blue Collar Engineering, a Louisiana-based company, gives an example of how AI can help engineers pick the best materials for a project. The firm explains that choosing the best-performing and most cost-effective materials is often a major challenge for engineers, but "AI can analyze vast amounts of data on material properties and recommend the most efficient materials for specific applications. By optimizing material usage during the design phase, engineers can reduce waste, lower production costs, and ensure that the final product meets performance requirements." It concludes, "AI's ability to quickly analyze multiple design variables allows engineers to strike the perfect balance between durability, cost, and efficiency."[56]

Automation and Robotics

Automation and robotics technology is also likely to be an important focus of mechanical engineering in the future. These technologies are advancing rapidly and expanding into many different fields, replacing humans in tasks that are tedious, are dangerous, or require a high level of precision. For example, robot vacuums clean homes, robotic systems assemble products in factories, surgical robots perform precise operations, and other robots are used in space exploration. Mechanical engineers design and repair these robotics systems, ensuring that they are both safe and effective.

Madi Babaiasl works in robotics engineering, and much of her work has focused on medical robots. She says, "During my robotics career, I worked on a variety of projects that I believe impacted society in general." For instance, she says, "My Ph.D.

research developed a new method for steerable needle insertion. . . . The steerable needle can turn inside the patient and get to difficult-to-reach places in the body to administer drugs or perform medical operations."[57]

Nanotechnology and Advanced Materials

Just as advances in robotics are transforming medicine and other industries, developments in nanotechnology are opening new possibilities for mechanical engineers to design and apply materials at the molecular level. This technology allows them to manipulate matter on an incredibly small scale and create innovative materials for a wide range of applications. The National Nanotechnology Initiative explains, "Using nanotechnology, materials can effectively be made stronger, lighter, more durable, more reactive, more sieve-like, or better electrical conductors, among many other traits."[58] Some current uses of this technology include developing flexible electronics that can be used in wearable devices and medical technology, fabrics that resist staining or wrinkling, and materials that can make cars lighter and more fuel efficient.

Experts believe that many more applications for nanotechnology and advanced materials will be developed in the future. For example, the National Nanotechnology Initiative points to new materials that can be used to charge phones and other devices. It says, "Researchers are developing thin-film solar electric panels that can be fitted onto computer cases and flexible piezoelectric nanowires [tiny crystalline wires that produce an electric charge when pressure is applied] woven into clothing to generate usable energy on the go from light, friction, and/or body heat to power mobile electronic devices."[59]

Additive Manufacturing

Also known as 3-D printing, additive manufacturing creates objects by depositing material layer by layer, following a digital blueprint. This technology was developed in the 1980s, but in recent years there have been advances both in the manufacturing process and the materials that it can produce. Everything from a tiny human blood vessel to a 25-foot-long (7.6 m) sailing vessel can be created using 3-D printers. Mechanical engineers are needed to help design these customized objects, and they also use this printing process to streamline their own work; for instance, by creating and modifying prototypes much more quickly and inexpensively and consolidating multiple printed parts into larger assemblies.

In an ASME article, writer Mark Crawford talks about how additive manufacturing is being used to produce affordable housing. He says, "A start-up firm called ICON recently 3D-printed a house for $10,000 using its Vulcan printer, which it claims is the first permitted and code-compliant home 3D-printed in the U.S. The company continues to advance its technology and hopes to reduce the total cost to $4,000 and the construction time to six hours."[60]

Mechanical Engineers Will Always Be Needed

In these fields and many others, there will always be a demand for mechanical engineers. These professionals not only recognize the need for society to continually change and improve,

A Digital Transformation

Mechanical engineer Kazuyoshi Fujimoto works at ULTMECHE, which provides résumé review and other types of career advice for mechanical engineers. He believes that there is a digital transformation occurring within this profession and insists that in the future, mechanical engineering roles will include a variety of tech skills. He maintains:

> The future of mechanical engineering is headed towards merging with tech. Hands down. Mechanical engineering skills will become integrated with coding, AI, and machine learning. There is no other way we can see the future. The world is shifting at a very fast pace. COVID-19 has accelerated remote working and leveraged the use of technology to maintain current business operations. . . . If you're an engineer in the industry, you should notice that companies are shifting towards a digital transformation.

Kazuyoshi Fujimoto, "The Future of Mechanical Engineering," ULTMECHE, February 29, 2024. https://ultmeche.com.

"My life's work is to create world-changing products that I thought were impossible. . . . The word 'impossible' is invigorating to me."[61]

—Malcolm Gutenburg, thermal engineer at NVIDIA

they frequently lead this change and expand the range of what is possible for humanity. Thermal engineer Malcolm Gutenburg works at NVIDIA, a company that makes computer chips and systems. He explains that mechanical engineers never see something as impossible. Instead, he says, "my life's work is to create world-changing products that I thought were impossible. . . . The word 'impossible' is invigorating to me."[61] This mindset is one of the things that makes this profession so valuable to society.

SOURCE NOTES

Introduction: Creating a Better Future

1. Quoted in Career Village, "What Do You Like/Dislike About Being a Mechanical Engineer?," June 29, 2023. www.careervillage.org.
2. UBC Engineering, "4 Reasons Why Engineering Is a Profession for the Future." https://engineering.ubc.ca.
3. American Society of Mechanical Engineers, "Ethics," 2024. www.asme.org.
4. Wassee Syed, "Ten Things I Wish I Knew in College—Tips for Mechanical Engineers," FEA Tips, October 11, 2022. https://featips.com.
5. Yassmin Abdel-Magied, "Yassmin Abdel-Magied, AMIMechE Well Engineer, Shell," Institution of Mechanical Engineers. www.imeche.org.

Chapter One: What Does a Mechanical Engineer Do?

6. Institution of Mechanical Engineers, "What Is Mechanical Engineering?" www.imeche.org.
7. SGW Team, "What Is the Engineering Design Process? An In-Depth Guide to Crafting Innovative Solutions," SGW Designworks, May 16, 2025. www.sgwdesignworks.com.
8. Quoted in Shelbi Tippett, "ME Careers: UMBC Alumni," Department of Mechanical Engineering, University of Maryland, Baltimore County. https://me.umbc.edu.
9. Quoted in IPG Automotive, "Aiming to Realize a Society with Zero Traffic Accidents Where Everyone Can Move Safely and Confidently," 2025. www.ipg-automotive.com.
10. Quoted in Haley Hirai, "How an NVIDIA Thermal Engineer Turns Up the Heat on Product Innovation," NVIDIA, February 24, 2025. https://blogs.nvidia.com.
11. Quoted in Lola Ravid, "Aerospace Engineer on How to Succeed as a Mechanical Engineer," Boundless Brilliance, August 1, 2022. https://boundlessbrilliance.org.

12. Kalifa Clarke, "A Day in the Life of a Biomedical Engineer: Kalifa Clarke," *All Together* (blog), Society of Women Engineers, December 1, 2021. https://alltogether.swe.org.
13. Quoted in Nat Levy, "Gabriel Cossio Is Making Nanotechnology Accessible for All," University of Texas at Austin Cockerel School of Engineering, October 2, 2024. https://cockrell.utexas.edu.

Chapter Two: How Do You Become a Mechanical Engineer?

14. Quoted in University of Michigan Mechanical Engineering, "Share Your Story," 2025. https://me.engin.umich.edu.
15. Quoted in University of Michigan Mechanical Engineering, "Share Your Story."
16. Syed, "Ten Things I Wish I Knew in College."
17. Peyton Erickson, "A Senior's Reflection on Mechanical Engineering," USC Viterbi Admission Student Engagement, February 9, 2024. https://viterbiadmission.usc.edu.
18. Quoted in Tippett, "ME Careers."
19. Quoted in Tippett, "ME Careers."
20. Quoted in University of Michigan Mechanical Engineering, "Share Your Story."
21. Quoted in Melissa F. Priebe, "Space Physics Q&A: An Interview with Chief Mechanical Engineer Jon Van Noord," University of Michigan Climate and Space Sciences and Engineering, October 20, 2022. https://clasp.engin.umich.edu.
22. Brian Calbeck, "My Path to Google: Brian Calbeck, Mechanical Manufacturing Engineer," *Life at Google* (blog), Google, May 29, 2019. https://blog.google.
23. Quoted in *U.S. News & World Report*, "Mechanical Engineer Overview." https://money.usnews.com.
24. American Society of Mechanical Engineers, "The Power of Lifelong Learning for Mechanical Engineers," June 22, 2023. www.asme.org.
25. American Society of Mechanical Engineers, "Ethics."

Chapter Three: What Skills and Personality Qualities Matter Most—and Why?

26. Chartered Institute of Personnel and Development, "Skills Matching: Using and Deploying People's Skills Effectively in the Workplace," December 9, 2024. www.cipd.org.

27. Quoted in Jacob Imm, "Is Mechanical Engineering Hard?," North Central College, December 3, 2021. www.northcentralcollege.edu.
28. People of Pacific, "What Can You Do with a Mechanical Engineering Degree?," University of the Pacific, February 6, 2025. www.pacific.edu.
29. Discover Engineering, "The Role of Physics in Mechanical Engineering," 2025. www.discoverengineering.org.
30. Quoted in Engineer Girl, "Featured Question: What Do You Love About Being an Engineer?," December 19, 2022. www.engineergirl.org.
31. Michigan Tech College of Engineering, "What Is an Engineer?," 2025. www.mtu.edu.
32. Pallavi Baddam, "Advice for Young Engineers: Be Open-Minded and Keep Learning," *Forbes*, January 26, 2021. www.forbes.com.
33. Meagan, "10 Things Aspiring Engineers Should Know How to Do," Engineering Expectations, 2025. https://engineeringexpectations.com.
34. Yee Siow, "What I've Learned in My Four Years in Mechanical Engineering," *Student Blog*, Imperial, May 31, 2022. https://blogs.imperial.ac.uk.
35. Discover Engineering, "Professional Tips for Mechanical Engineering Success," 2025. www.discoverengineering.org.
36. Quoted in Engineer Girl, "Do You Have to Be Really Outgoing to Be an Engineer?," December 21, 2018. www.engineergirl.org.

Chapter Four: What Is It like to Work as a Mechanical Engineer?

37. Yassmin Abdel-Magied, "Meet a Professional: Yassmin Abdel-Magied AMIMechE, Well Engineer, Shell," Institution of Mechanical Engineers. www.imeche.org.
38. Institution of Mechanical Engineers, "Where Do Mechanical Engineers Work?" www.imeche.org.
39. Quoted in *Career Cultivation* (blog), AgCareers.com, "What It's Really like to Be a Mechanical Engineer," July 18, 2017. https://blog.agcareers.com.
40. Quoted in *Career Cultivation* (blog), AgCareers.com, "What It's Really like to Be a Mechanical Engineer."
41. Quoted in Axis Communications, "Meet Malin, Mechanical Engineer and Nature Lover," May 9, 2023. https://lifeat.axis.com.

42. Brandon Hayter, "My Daily Life as an Engineer," Peak Life, August 23, 2022. https://peaklifebrandon.com.
43. Ben Truman, "Meet a Professional: Ben Truman AMIMechE, Global Engineering Lead—Actuation & Propeller Systems, UTC Aerospace Systems," Institution of Mechanical Engineers. www.imeche.org.
44. Quoted in *CME Blog*, Custom Manufacturing & Engineering, "Interview with a Mechanical Engineer I: Erin," March 4, 2021. www.custom-mfg-eng.com.

Chapter Five: Advancement and Other Job Opportunities

45. Quoted in Adam Calica, "A Day in the Life of 29 Engineering Managers," Built In, July 13, 2023. https://builtin.com.
46. Quoted in MECAGINE, "Mechanical Engineering Manager: Nathan's Testimonial." www.mecagine.com.
47. Cathy Cecere, "Engineering Jobs That Are Both Ordinary and Extraordinary," American Society of Mechanical Engineers, January 11, 2024. www.asme.org.
48. Truman, "Meet a Professional."
49. UC Riverside, "12 Alternative Careers for Engineers." https://engineeringonline.ucr.edu.
50. Emilie, "My Career Journey from Motorcycle Mechanic to Mechanical Engineer," *The Female Engineer: Stem Lifestyle Blog*, April 21, 2021. https://thefemaleengineerblog.com.
51. Emilie, "About Me," *The Female Engineer: Stem Lifestyle Blog*, 2023. https://thefemaleengineerblog.com.
52. Mary Frecker et al., "What Is Mechanical Engineering?," *Mechanical Engineering Magazine*, July 2024. https://asmedigitalcollection.asme.org.

Chapter Six: What Does the Future Hold for Mechanical Engineers?

53. Cathy Cecere, "When It Comes to Engineers, Demand Exceeds Supply," American Society of Mechanical Engineers, April 16, 2025. www.asme.org.
54. Quoted in *U.S. News & World Report*, "Mechanical Engineer Overview."

55. Jerry Guerra, "Mechanical Engineers Thrive in the Green Economy," American Society of Mechanical Engineers, June 4, 2025. www.asme.org.
56. Blue Collar Engineering, "From Design to Production: How AI Is Changing Mechanical Engineering," October 2, 2024. www.bluecollar.engineering.
57. Madi Babaiasl, "A Day in the Life of Robotics Engineer Madi Babaiasl," *All Together* (blog), Society of Women Engineers, April 29, 2022. https://alltogether.swe.org.
58. National Nanotechnology Initiative, "Applications of Nanotechnology." www.nano.gov.
59. National Nanotechnology Initiative, "Applications of Nanotechnology."
60. Mark Crawford, "5 Creative Applications of Additive Manufacturing," American Society of Mechanical Engineers, May 9, 2023. www.asme.org.
61. Quoted in Hirai, "How an NVIDIA Thermal Engineer Turns Up the Heat on Product Innovation."

Scott Bohn is an engineer who works in Boise, Idaho. His career has included working in the mechanical engineering field. He answered questions about his career by email.

Q: What kind of work do you do?

A: I oversee the companies' criteria specifications for refrigeration and HVAC [heating, ventilation, and air-conditioning] equipment. [These are the standards and features that ensure that this equipment is safe and efficient.] In addition, my team manages facility refrigeration projects across our entire store portfolio. I have been working in the retail grocery industry for a little over 6 years now. I began working in the mechanical engineering field at the company during my time in equipment procurement, where I first began establishing our criteria and specifications for our companies' refrigeration system designs. My initial education was in electrical engineering and computer science, which I applied for 5 years after graduation, in the semi-conductor industry.

Q: Why did you become an engineer?

A: I went back to school for my electrical engineering degree after I had already begun a career as a journeymen electrician [a licensed electrician who has completed an apprenticeship]. This is where I first started learning and applying some of the basic concepts related to several of the basic engineering principles. After some time in the trade, I decided I wanted to have a greater understanding of the principles behind what I was implementing in the field. I felt that having an engineering degree would be a valuable asset to companies while still allowing me the opportunity to remain in the construction industry I so much enjoyed.

Q: What is your typical workday like?

A: A typical work day for me centers around managing refrigeration upgrade or replacement projects. My team replaces outdated refrigeration equipment with new, efficient, equipment that has less impact on the environment. Whether that is working with my internal team to review potential future projects, working with our design engineers on planning current projects, or commissioning finalized projects to ensure that they meet our companies' specifications and that the scope of work has been completed.

Q: What do you like most about your job?

A: I would say that the best part of my job is that I get to continue growing and learning in so many areas outside of my specific area of focus. When you have an engineering degree you are able to apply those problem solving skills towards so many career paths. While I was trained in the area of electrical engineering, I feel the most important skill learned in all of the engineering fields is how to problem solve and implement a strategy. These skills have allowed me to work in an adjacent engineering field to my education and be successful. I think that is why engineers from all practices are so valuable to companies. Employers understand that engineers are problem solvers and that skill is extremely valuable to all industries.

Q: What do you like least about your job?

A: While the analysis and design elements of what I do are something that I truly enjoy, it limits my opportunity to be out in the field seeing practical implications of the equipment. Troubleshooting was always something I loved when I was in the field, but in my current role, there is a limited amount of time that I am able to spend in the field.

Q: Is there anything about your job that is not the way you expected it to be?

A: Engineers are often thought of, I feel, as isolated from the rest of the company, just designing and sending out what is to be done. I think that this perception couldn't be further from the truth. Oftentimes the most successful engineers that I have encountered are the ones that have great people skills. Being able to communicate with customers and other areas of the business so that everyone has a clear understanding of what you are trying to accomplish is often as important as the design itself.

Q: What personal qualities do you find most valuable for this type of work?

A: I think the most important qualities anyone in the engineering industry can have are the ability to problem solve and a real desire to learn and grow. Industries that require engineers never sit still and neither can the people working in them.

Q: What is the best way to prepare for this type of engineering job?

A: You need to understand the equipment and its relationship to the refrigeration lifecycle. Understanding the relationships between temperatures and pressures and how outside factors can be used or altered to reach the desired effect of the refrigeration cycle to meet your systems needs. A solid understanding of the basic principles of thermodynamics is a must.

Q: What advice do you have for students who might be interested in a career as an engineer?

A: Don't get discouraged when things get difficult. When you are able to understand or figure out the solution to something you thought you never could, that is when the job is the most rewarding.

American Society for Engineering Education (ASEE)
www.asee.org
The ASEE publishes numerous journals about engineering. Its website also has a variety of educational resources, including profiles of engineering colleges in the United States, details about community college paths to an engineering degree, and demographic information about engineering education schools and educators.

American Society of Mechanical Engineers (ASME)
www.asme.org
The ASME website contains articles on a wide range of topics related to mechanical engineering, including information about different types of mechanical engineering jobs and new trends and technology in this field. It also has career advice, such as tips on developing an effective job application and identifying positions that align with a candidate's skills and aspirations.

Institution of Mechanical Engineers
www.imeche.org
This organization's website has educational information about mechanical engineering, including details about what mechanical engineers do and the many different industries and specializations available to them. It also has a variety of articles about new developments in this field.

Society of Women Engineers
https://swe.org
The Society of Women Engineers publishes information about women in engineering, including information about representation and the top engineering fields for women. It also publishes the yearly *State of Women in Engineering* magazine.

INDEX

Note: Boldface page numbers indicate illustrations.

PICTURE CREDITS

Cover: Patamaporn Umnahanant/iStock

4: Maury Aaseng
5: Shutterstock
10: Nutchapong Wuttisak/Shutterstock
19: JK1991/iStock
30: visualspace/iStock
38: Associated Press
43: eakgrunge/iStock

ABOUT THE AUTHOR

Andrea C. Nakaya holds a bachelor of arts in English and a master of arts in communications from San Diego State University. She has written and edited hundreds of articles and more than fifty books on current issues. She currently lives in Eagle, Idaho.